THE MICROBE FILES

Cases in Microbiology for the Undergraduate

Marjorie Kelly Cowan
Miami University of Ohio

Benjamin
Cummings

San Francisco   Boston   New York
Capetown   Hong Kong   London   Madrid   Mexico City
Montreal   Munich   Paris   Singapore   Sydney   Tokyo   Toronto

Publisher: Daryl Fox

Sponsoring Editor: Amy Folsom Teeling

Associate Project Editor: Erin Joyce

Managing Editor: Wendy Earl

Production Supervisor: Sharon Montooth

Cover and Interior Design: tani hasegawa

Copyeditor: Anita Wagner

Compositor: The Left Coast Group, Inc.

Manufacturing Supervisor: Stacey Weinberger

Marketing Manager: Lauren Harp

Library of Congress Cataloging-in-Publication Data

Cowan, M. Kelly.
    The microbe files : cases in microbiology for the undergraduate / M. Kelly Cowan--1st ed.
      p. cm.
    Includes index.
    ISBN 0-8053-4928-6
     1. Medical microbiology--Case studies. I. Title.

 QR46.C866 2001
 616'.01--dc21

                                                  2001017162

ISBN: 0-8053-4928-6

26 25 24 23 22 21 -RRD- 15 14 13

Dedication

This book is dedicated with love and gratitude to my sons and my husband.

About the Author

Marjorie Kelly Cowan, Ph.D., teaches microbiology and epidemiology at the Middletown campus of Miami University in Ohio. Her training took place at the University of Louisville, the University of Groningen (the Netherlands), and the University of Maryland. She actively studies teaching and learning, and oversees a research program in microbial adhesion and colonization in medical and industrial settings. She recently received a Celebration of Teaching Award from the Greater Cincinnati Consortium of Colleges and Universities. She has two sons, Taylor, 12, and Sam, 9.

Acknowledgments

Thank you to my students, who continuously teach me how to teach and show me why it is worth doing. Many of the cases in this book originated in stories students told me about their own experiences in health care.

Of course many people contributed in direct and indirect ways to this project. The book came about through the vision of Amy Folsom Teeling (Sponsoring Editor) and the hard work of Erin Joyce (Associate Project Editor) at Benjamin Cummings. Jeff Howard came up with the title. Sharon Montooth shepherded the book through production. Thanks also to Jim Lewellyn and Suzanne Evans at Miami University for proofreading and assistance. I want to publicly acknowledge the leadership and support of Anne Morris Hooke, chair of microbiology at Miami University. Her humanity and dedication to excellence, fairness, and integrity, are an inspiration to all who know her.

This book was much improved by the comments and suggestions of a multitude of reviewers. Their passion for teaching was evident in their thoughtful reading and commentary, and I have deep appreciation for their input: Cindy Anderson, Mt. San Antonio Community College; Rick Bliss, Yuba College; Burton Clark, Oregon Institute of Technology; Diane Dudzinski, Washington State Community College; Christine Frazier, Southeast Missouri State; Alan Gillen, Pensacola Christian College; Hank Harris, Iowa State University; Leslie Lichtenstein, Massasoit Community College; Katie Morrison-Graham, Lane Community College; Carmen Rexach-Zellhoefer, Merced College; Leba Sarkis, Aims Community College; Phillip Taylor, Glenville State College; Karin VanMeter, Des Moines Area Community College; Ruth Wrightsman, Saddleback College; and Susan Wyckoff, Bradley University.

Finally, heartfelt thanks to my husband Paul Wehner. What a world it would be if everyone had just one person like Paul to believe in them.

Marjorie Kelly Cowan

Contents

Introduction

Dear Student,

When you are learning about infectious diseases, the amount of information you have to master may seem overwhelming. There are so many infections, so many microorganisms, so much new terminology. This book is designed to help your brain organize those long lists of facts into logical pathways. You'll be solving cases instead of rehashing the same old facts. Once you've placed a piece of information into the right place in the big picture, it will have a meaning for you that will be almost impossible to forget!

The Microbe Files

The six chapters in this book provide examples of infectious diseases presented to you the way you will encounter them in your life or clinical practice. The cases are grouped by chapter according to the body site that is most affected. This mirrors what happens in clinical practice: your first encounter with a patient's infection will be what you can see and what the patient can tell you.

In the News Cases

You will find an "In the News" case in every chapter. If you've picked up a newspaper since you've started studying microbiology you may have noticed that nearly every day you can find an article about a new infection, a new epidemic, or a new treatment for a disease. And now you can read these with a knowing eye! The "In the News" cases in this book will get you started in applying your expertise.

Challenge Cases

Each chapter also contains at least one "Challenge" case. Like the other cases in the book, these should require no further information than that obtained in your course. They will ask you to push yourself and to apply what you know within a new context or with a different twist.

The Usual Suspects

Each chapter contains a box that lists the most common infectious agents causing symptoms in the body system under consideration. Don't refer to the list right away; but if you feel stumped, flip to it to see if there's something you missed in your mental browsing.

Glossary

You will also find a glossary in the back of the book that contains many terms that are either very important or particularly mysterious. It is easy to forget the glossary is there, so to remind you to use it, each glossary word appears in **boldface** type the first time it appears in a case.

Steps in Diagnosing a Patient

Anatomical Diagnosis

Clinical diagnosis has a pattern, and typically it goes something like this: a patient comes to you with obvious signs or symptoms, or has a verbal explanation of invisible body sensations. (This is called the *presenting* complaint or symptom.) He or she wants a diagnosis. Even though we have enormous amounts of diagnostic technology available to us, it is obviously impractical (not to mention unethical!) to give patients every available diagnostic test every time they experience illness of any type. The inevitable conclusion is that reaching a diagnosis requires complex critical skills on the part of the practitioner (you) to determine which, if any, diagnostic tests must be performed. Right away, the presenting symptoms allow you to narrow the possibilities. The symptoms will most likely affect a particular area of the body most obviously—like the skin, or the digestive tract. At this point your list of hundreds of possible infections becomes a list of a handful of possible infections. It should be noted that one whole category of diseases seems to affect the entire body. These infections are called **systemic,** and once again, knowing that will help you narrow your choices. Identifying the physical site of symptoms is called the **anatomical diagnosis** and is the first step in clinical diagnosis.

Differential Diagnosis

At this point you have a short list of possible diseases—infectious and noninfectious—that may be causing the symptoms. The list of diseases that could be causing the anatomical symptoms is called the **differential diagnosis.** The differential diagnosis may contain only two or three diseases. If the presenting symptom in a 5-year-old patient is a sore throat with no other symptoms, *Streptococcus pyogenes* or viruses (as a group) are probably the only two causes in the differential diagnosis. On the other hand, a single differential diagnosis may list several infectious agents plus a few metabolic disorders. Consider the symptom of jaundice: it could be caused by one of several hepatitis viruses (which must be sorted out) as well as alcoholism or inherited disorders. On the other end of the spectrum, in a few infections the outward signs are so unique to a particular microorganism that only one infectious cause is possible. The bull's-eye rash of Lyme disease, when present, is diagnostic of *Borrelia burgdorferii* infection, for example.

Etiological Diagnosis

Once you've got the list narrowed to a differential diagnosis, it is time to make the **etiological diagnosis.** In this phase, the actual causative organism is identified, if it is an infectious disease. In the case of jaundice, blood would be drawn and immunologic tests conducted for each of the hepatitis viruses. Other causes can be investigated with blood samples as well (by examining levels of liver enzymes, etc). Sometimes culture of the infected site must be performed.

Epidemiology

Midway through the process of generating the differential diagnosis, and continuing into the etiological diagnosis phase, **epidemiological** considerations become indispensable. This means that you need to ask the following questions related to diseases you are going to list in the differential diagnosis: What people are at risk for this disease? Do particular behaviors (traveling, smoking, etc.) expose one to the disease? (Has this patient engaged in these behaviors?) What is the geographical distribution of the disease? (Has the patient been in these locales?) Have I seen a lot of this disease recently in my practice? Has the patient been immunized against this disease? (Is there a possibility for vaccine failure?) Of course, the answers to many of these questions are obtained from the patients themselves, and clinicians have to be discerning about the patient's ability to "self-report." So in addition to being an epidemiologist you must be a bit of a psychologist as well.

Diagnosis in Practice

In practice, the first two phases—anatomical and differential diagnosis—occur very quickly. The third phase often takes time and money, which is reason enough to make good guesses in the earlier phases. It is important to note that in practice the third phase is often not necessary at all. This occurs for several reasons. If all of the possible etiological agents would be treated in the same way, or not at all (viral rhinitis, for instance), there is no need to identify the organism. Also, the decision not to identify the organism can be determined by a delicate balance between the need to know and the degree of discomfort or expense to the patient. A very common example of this last situation is the pediatric ear infection (otitis media). Typically the physician prescribes treatment solely on the basis of clinical signs, even though several different bacteria can cause the infection. Obtaining samples from the middle ear for culture is unpleasant for the patient, and in the past physicians found that simply prescribing a broad-spectrum antibiotic was the best approach for handling these infections. Now that there is great concern about antibiotic resistance, physicians often opt for "watchful waiting" instead of prescribing an antibiotic in the case of ear infections. But even in this scenario, an etiological diagnosis is usually not made. Ear infections of all types usually resolve themselves within a few days.

For some common conditions, written flow diagrams exist for making all of these decisions and treating the patient ("If this symptom is present, perform the following tests"). These are sometimes called algorithms, or clinical pathways, and may be written by hospital administration and be part of hospital policy. Insurance companies also have diagnosis and treatment algorithms. But it is difficult to predict all possible scenarios, and most of the time diagnosis still requires your own good judgment and experience. Accordingly, some of the cases in this book illustrate what you, as a student of the health sciences and of biology, already know. Health, because it is a biological phenomenon, has a lot of gray areas. There is no single rule of diagnosis. The pattern of diagnosis described here is just that—a pattern. You will diverge from that pattern when your clinical judgment tells you it is appropriate. The cases in this book are designed to help you learn the pattern, and learn how to determine when to veer from it.

Practice Flying

I tell my students that exams cause anxiety because they ask you to fly solo—to answer questions with a minimum of background information available to you. Therefore, I tell them, you have to *practice flying* while you study. Pose problems to yourself without giving yourself access to the answers. Don't give up at the first twinge of difficulty and flip to the answer in your notes, or in the back of the book. Use this opportunity while you're still on the ground to push yourself further. Experience the discomfort of racking your brain now, and later, when it counts, you will be less likely to have to. It will pay off on the exam, and more importantly, it will pay off in your practice. You can think of this book as your flight simulator!

Best of luck,
Marjorie Kelly Cowan

THE
MICROBE
FILES

Diseases of the skin and Eyes

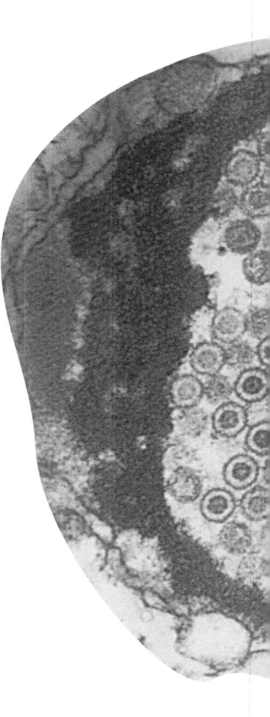

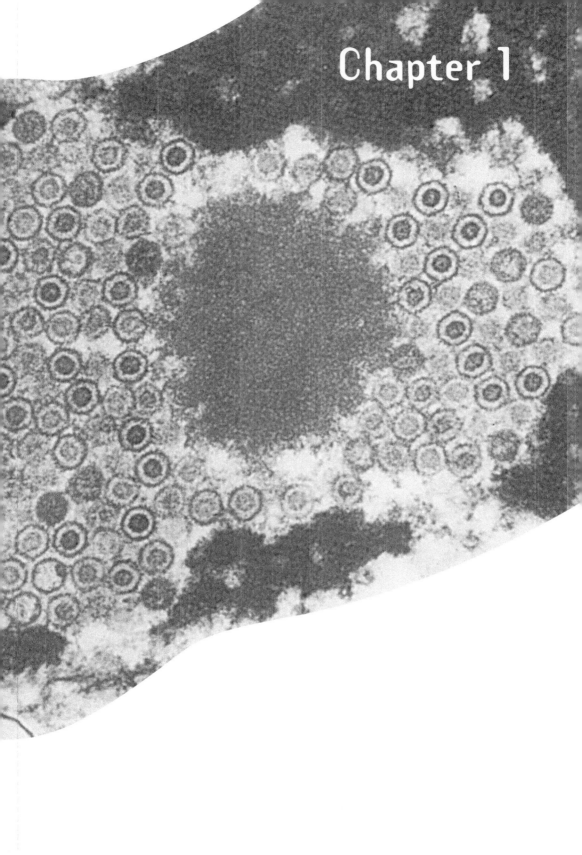

Chapter 1

Chapter Opener One is a highly magnified photo of *Herpesvirus.*

Case 1.1

Kate, your sister-in-law, is about to undergo fertility treatments. Her doctor insists that she receive the rubella vaccination, and then wait several weeks before beginning the actual fertility regimen. Kate calls you and wants to know why she has to do this. You ask her if she is able to produce evidence of vaccination for rubella (also known as German measles). She says no; her family had a house fire a few years ago and all those records were lost.

"But I had German measles when I was in second grade!" she says. "I remember that I was really sick and missed almost a month of school."

You suggest that she follow her doctor's advice and get the immunization.

1. Why would a fertility specialist recommend the rubella vaccine? Why does he suggest a waiting period after vaccination and before conceiving?

2. When do most children in the United States receive their rubella immunization?

3. Kate suggests that she had rubella in second grade, but the disease she described doesn't sound like rubella to you. Why not?

4. Kate says the doctor gave her the option of having her blood checked for antibodies to the virus, to test her immune status. Would this test be checking for immunoglobulin M or G (IgM or IgG)? Explain your answer.

5. If a physician was checking for a current rubella infection and only had available a test for IgG, how could he or she be certain the infection was a new one?

Case 1.2

In late September a woman brings her 14-year-old daughter, Meg, to the family physician. Meg shows the doctor the back of her thigh where there are pale red, nonraised discolorations. The rash covers a wide area of the thigh and seems to be roughly circular. The center of the circular area appears normal. Meg has no other symptoms, but her mother brought her in because the rash has been present for over three weeks and it seems to be growing.

The doctor questions Meg about possible exposures. Has she worn any new pants lately? Has she been in the woods? Do her joints hurt? Meg reports that she spent the month of August at summer camp in Vermont. She's been wearing mostly shorts and bathing suits for the past two months, none of them new. She doesn't remember any insect bites on her thigh.

1. On the basis of Meg's **oral history,** what is the most likely diagnosis? What would the causative microorganism look like in a Gram stain?

2. How did she most likely acquire her infection?

3. Would the diagnosis be any different if Meg had attended camp in Arizona? Explain.

4. Why does the doctor ask Meg if her joints hurt?

5. How is this infection treated?

6. Meg's mom, upon hearing the **presumptive diagnosis,** declares that Meg will not return to that camp, which she loves and had planned to attend next summer. The doctor suggests that Meg need only take some precautions. How can she protect herself from getting this infection again?

Case 1.3

You have a possible infectious condition that you are embarrassed to discuss with the physician with whom you work. You have worn artificial nails for several months now and noticed that the one on your left ring finger falls off regularly. The real nail underneath has become white and chalky, and the skin around the nail is beginning to have little white lines in it and look a bit chalky, as well.

1. What disease do you suspect? Explain why.

2. What would you suggest be done for a more definitive diagnosis?

3. Can you treat this yourself with an over-the-counter drug, or do you need to see a physician?

4. You see cures for this condition mentioned on TV and on the Internet—do you think they work?

5. What other conditions are caused by **dermatophytes** (*Microsporum, Trichophyton,* and *Epidermophyton*)? What is special about them that makes them capable of thriving in their anatomical **niche** on their hosts?

Case 1.4

You are the school nurse at Willowdale Elementary. This morning Ms. Matthews, one of the first-grade teachers, brings a little girl named Keisha to your office. Her right eye is swollen and bloodshot. The lining of the lower lid is bright red. There is a thick yellow discharge in the corner of the eye.

1. What is the most likely diagnosis, and what is the **etiology**?

2. What sign leads you to believe that the infection is bacterial in origin?

3. What is the treatment for this condition? Elaborate. Is the condition **communicable**?

4. What are some of the eye's natural defenses that help to prevent infections?

5. Are there steps the teacher should take to prevent the spread of this infection in the classroom? If so, discuss them.

Case 1.5

A woman is brought to the emergency department where you are working **triage.** She has an extremely swollen right lower leg. You see what appears to be an old surgical wound in the mid-calf, with rough scar tissue surrounded by purplish-red skin. She is in a lot of pain and her husband speaks for her. He tells you that three weeks ago she had a group of moles removed from that area. It had appeared to heal initially, but three days ago the incision area started looking bigger rather than smaller. She did not return to the physician, hoping the condition would resolve itself. In the past three days the area has begun to swell and become very hot.

You call the attending physician immediately because you know that this is a serious condition.

The patient is sent straight to surgery where the wound is **debrided.** Gram-positive cocci growing in chains are recovered from the wound. She is transferred to intensive care and put on high-dose intravenous antibiotics for the next 18 hours, but the next evening her leg is amputated below the knee. She remains in the hospital for two months following surgery and requires long-term antibiotic therapy and multiple skin grafts on her upper leg.

1. What condition did this patient have? What features suggest that it is not *Clostridium perfringens* gangrene?

2. Why was amputation the best solution for the infection in this case?

3. How is the bacterium transmitted?

4. It seems like we've heard a lot more about this condition in the past few years. Is this just media hype or are more cases occurring? Explain.

Case 1.6 In the News

In the late winter of 1988, pediatricians in big cities around the country started reporting large increases in the numbers of patients they saw with diffuse red rashes and high fevers (greater than 101°F). The rash, usually extending downward from the hairline to the rest of the body, began after a two-week incubation period. The spots were often so close together that the entire involved area appeared red. Sometimes the skin in such an area peeled after a few days. The rash lasted five to six days. Many of the children also suffered from diarrhea.

The age group most affected was preschoolers. This was a change in **epidemiology** for this infection, as previously the disease most often struck school-age children. A vaccine had been introduced for this disease in 1963, and since then only 5000–6000 cases a year had been reported in the United States. In 1989, 18,193 cases were reported. In 1990 the **epidemic** peaked with almost 28,000 cases reported in the United States. Since then the incidence in this country has fallen rapidly and is again in the range of 5000–6000 cases a year.

1. What was this resurgent infection?

2. What are some possible reasons for the epidemic in 1989–1991?

3. What is **herd immunity**? Discuss it in relation to this outbreak.

4. What is the schedule for vaccination for this infection in the United States?

5. Are serious **sequelae** associated with this infection? If so, what are they?

Case 1.7 Challenge

A woman brings her 6-month-old son to the pediatrician. You are following the doctor as part of your physician's assistant training. Before you enter the examining room the physician pulls the chart off the door and hands it to you. The nurse has written on the chart that the chief complaint is a group of lesions on the child's back.

You enter the room and greet the mother. A toddler girl is leaning over to play with the baby in his carrier on the floor. The baby is giggling and appears healthy. You notice on the chart that the baby was breast-fed from birth through his fourth month. Mom explains that the spots on the baby's back just popped up two days ago and that the baby hasn't had a fever and seems well. She lifts the baby up and you examine the lesions—a group of about seven to eight blisterlike lesions **localized** to the left of the baby's spine. They have clear fluid in them. The physician says the lesions are diagnostic.

1. What are the lesions diagnostic of? Explain how you decided.

2. Although this particular condition is somewhat unusual in babies, the lesions indicate that the child must have experienced a common childhood illness earlier. Which one?

3. The mother says that, to her knowledge, the baby has not had this common childhood illness, but that his 3-year-old sister had it four months ago, when the baby was 2 months old. Explain the link between the girl's illness and the baby's condition.

4. What factors probably influenced the fact that the baby did not have symptomatic illness when his sister was experiencing it? And what factors led to the eruption of lesions now?

5. Is this a dangerous condition? Why or why not?

Case 1.8 Challenge

Your stepbrother John is 5 years old. One day he comes to the breakfast table with a bright red face, almost as if he had been slapped. When you look more closely you can see thousands of tiny red bumps on the skin. He has a milder rash on his arms and legs and just a few red bumps on his trunk. He isn't acting sick and doesn't have a fever. He had chicken pox when he was 3 and his immunization schedule is up to date. He sticks his tongue out at you while you're examining his skin and that reminds you to check his throat, which looks normal, no redness. He says his throat hasn't felt sore. Your mom mentions that he has had a runny nose for the last few days, but he hasn't felt ill.

1. Your diagnosis? Why was his throat checked?

2. Can John go to kindergarten today? Why or why not?

3. Is this infection rare? Explain.

4. Are any **sequelae** associated with this infection? If so, name them.

The Usual Suspects

Common microorganisms causing infections on the skin or in the eyes[1,2]

Bacteria

Gram-positive

Propionibacterium species
Pseudomonas aeruginosa
Staphylococcus aureus
Streptococcus pyogenes

Gram-negative

Borrelia burgdorferi
Chlamydia trachomatis
Haemophilus aegyptius
Mycobacterium leprae
Neisseria gonorrhoeae

Viruses

Adenovirus
Coxsackievirus
Echovirus
Herpes simplex virus
Measles virus
Papillomavirus
Human parvovirus B19
Rubella virus
Varicella-zoster virus

Fungi

Candida albicans
Epidermophyton
Microsporum
Sporothrix schenckii
Trichophyton

Protozoa

Leishmania species
Loa loa
Onchocerca volvulus

[1] Not all of the infections appear in this chapter.
[2] Not an exhaustive list.

Diseases
of the
Nervous
System

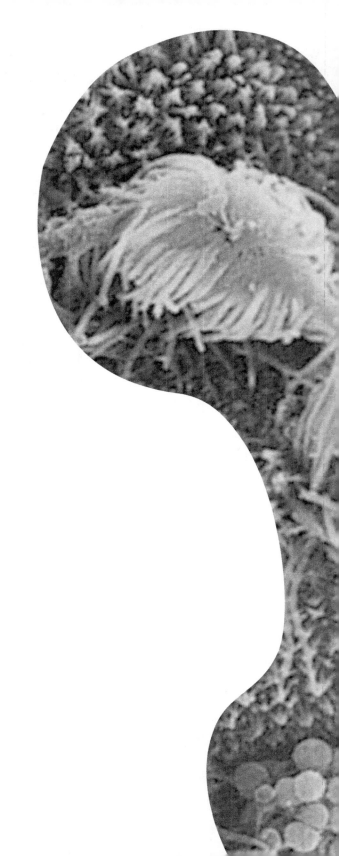

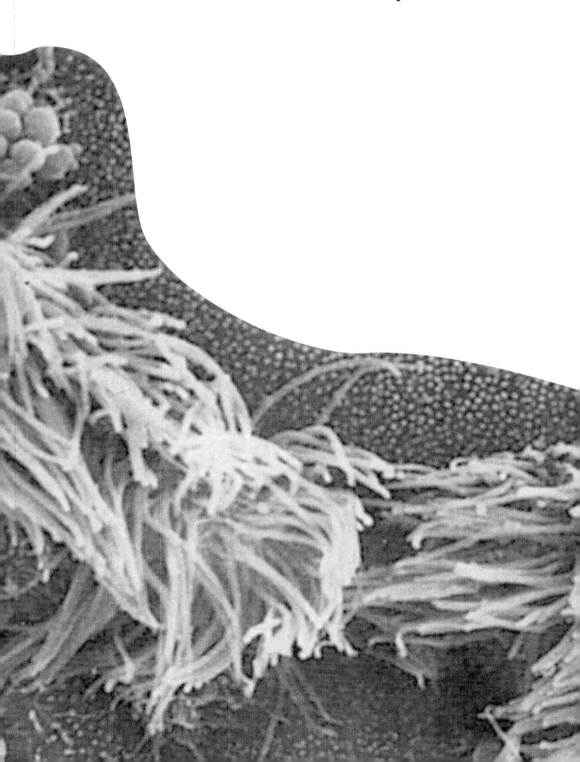

Chapter Opener Two is a highly magnified photo of *Neisseria meningitis*. Bacteria are the coccoid structures near top and bottom of photo.
Courtesy of D.S. Stephens, Emory University School of Medicine.

Case 2.1

A mother brings her baby daughter in to your office for her 12-month set of vaccinations. The baby is scheduled to receive, among others, the **MMR** and the polio immunizations. After the nurse shows the mother, the baby, and the baby's 3-year-old brother to the examining room, she tells the mother to undress the baby except for her diaper. She hands her a blue booklet about the polio vaccine and additional forms about the other vaccines and asks her to read them all.

The baby cries as the mother undresses her. The 3-year-old starts to climb up a chair and reaches for the needle-disposal container mounted on the wall. The young mother is frazzled; she keeps one hand on the baby lying on the table and tries to scoop up her son with the other. The 3-year-old slips out of her grip and bumps his knee on the corner of the table as he skids to the floor. Now he is crying, too.

The mother never manages to read the vaccination information but she does sign the forms where she is supposed to. When the physician arrives, she checks for the signatures, then performs a thorough examination. On her way out she tells the mother that the nurse will be in to administer the vaccinations.

1. What type of information is contained in all vaccine brochures? Why should they be read before the vaccines are administered?

2. What particular facts are critical for parents to know about the polio vaccine?

3. What other vaccines besides the MMR and polio are appropriate for a 1-year-old child?

4. Two forms of the polio vaccine are available, the live attenuated version (called the OPV, or oral polio vaccine) and the killed or inactivated version (called the IPV, or injectable polio vaccine). Why is the OPV the preferred version for this age?

Case 2.2

You are doing a rotation in the hospital's clinical laboratory. A sample of cloudy cerebrospinal fluid (CSF) from a suspected meningitis case arrives and you are told to Gram stain it, and then to plate it on blood agar and **chocolate agar.** In the Gram stain you find gram-negative rods of varying size and shape. You also find a lot of bacteria inside phagocytic cells. They are not diplococci. Colonies grow on both of the plates you inoculated.

Later, the charge nurse tells you that the patient, a 3-year-old girl, has not received any childhood vaccinations.

1. What is the most likely causative organism? Why?

2. Why was the child's unvaccinated status helpful in diagnosis?

3. What is causing the cloudiness in the CSF?

4. What other types of infections can this organism cause in children?

Case 2.3

You are working in the emergency department of a regional hospital in rural Kentucky. A patient is brought in by emergency medical technicians (EMTs). Their initial report is suspected meningitis because the patient has a headache and stiff neck. The EMTs add that the patient's meningitis symptoms appear rather mild—he still has neck movement, and the headaches are not severe. The patient's overall condition is poor, however. He is very thin, has dark spots on his face and upper body, and open bloody-looking eruptions on his lips. His fever is 104°F, and his blood pressure is low. He also has severe diarrhea.

1. What is the first step in determining if the patient has meningitis?

2. This test reveals the presence of very large cells that appear to be eukaryotic, surrounded by a large capsule. What is the probable diagnosis? Name some other eukaryotic organisms that can cause meningitis symptoms.

3. What groups of people are at risk for this infection?

4. How is it acquired?

5. What anatomical sites are most often infected with this fungus?

6. Let's say your initial suspicion (your answer to question 2) was correct. What other diagnostic test should be performed on this patient?

Case 2.4

You are an emergency medical technician and are called to the home of Kevin, a 13-week-old boy who has become listless and is having trouble breathing. The parents report that Kevin used to smile, but lately he has not smiled, nor has he had other noticeable facial expressions in the last two days. Kevin's eyes are open when you arrive, but he does not seem to be focusing. You place your out-stretched finger under his fingers and he fails to grasp it. You lift his foot and it drops back to the mattress. The parents report that he has not had a bowel movement in three to four days.

1. What is your suspicion, based on what seem to be nervous system symptoms?

2. If this is indeed the case, do you start treatment here at Kevin's home, or should you transport him to the local hospital?

3. What should be administered to Kevin at the earliest opportunity?

4. How do babies acquire this condition?

5. Although the diagnosis should be confirmed with laboratory tests, the tests should probably not be performed in the hospital lab. Why not?

Case 2.5 In the News

In the winter of 1993, five students from one middle school in Seattle were diagnosed with meningococcal disease. The incidence of the disease had been climbing for two years in that area of Washington State, as well as in the rest of the country, and has continued to climb since then. In the Seattle outbreak, health officials identified one strain of the causative organism that was responsible for the increased incidence.

1. What type of organism would you look for in a Gram stain of blood or cerebrospinal fluid in these cases?

2. What is the organism's **portal of entry** to the host?

3. Could you swab the portal of entry (see question 2) to detect the presence of the organism? Why or why not?

4. What types of symptoms are associated with meningococcal disease?

5. A total of 900 students attend the affected middle school. What measures should have been taken to protect the remaining 895 students from acquiring meningococcal disease?

Case 2.6 In the News

In late July of 2000 the most famous park in the United States, Central Park in New York City, was closed to the public so that it could be sprayed with insecticide to prevent the spread of the West Nile virus. Parks department workers handed out pamphlets titled *Public Health Alert.*

West Nile virus had first been noticed in New York the previous summer and fall. Seven people were killed in that early outbreak, and 55 cases of the illness were confirmed. It had not been seen previously in the United States. As its name suggests, it is normally found in Africa, the Middle East, and western Asia, as well as in parts of Europe.

1. How is West Nile virus transmitted?

2. The virus had another vertebrate host besides humans when it showed up in New York. What was it?

3. Can you list some possible mechanisms for how the virus was introduced into the United States?

4. Most infections result in no noticeable symptoms. Some of those infected may develop a skin rash. A fraction of people infected develop life-threatening encephalitis. What is encephalitis and who do you suppose is most likely to experience this symptom?

5. A sudden increase in a particular disease within a population of humans is called an **epidemic.** What is a large outbreak among animals called?

6. If you lived near Central Park and wanted to go jogging there, what would be the best time of day to avoid the park to minimize your chances of being infected with West Nile virus?

Case 2.7 In the News

In the late 1980s there was an **epidemic** among livestock in Great Britain. Approximately 180,000 cattle were found to have "mad cow disease," so named because the condition attacks the central nervous system, which leads to bizarre behavioral symptoms and often death. The disease seems to be caused by an "unconventional transmissible agent," meaning it is unlike most microorganisms. No genetic material from this organism has ever been detected in infected tissues, although foreign protein fibers accumulate in large concentrations in the brain. Infected cows were still turning up in the late 1990s.

There was great alarm in the late 1990s when dozens of humans starting turning up with symptoms similar to those seen in the cows in the late 1980s. More than 50 people have been diagnosed with the human variant of mad cow disease. This is consistent with the approximately 10-year incubation period of this unconventional transmissible agent. In 1996 scientists confirmed that the same agent was present in affected human and cow brains.

1. What is the name for a transmissible agent that contains only protein and has no genetic material?

2. What is the formal name for mad cow disease? Explain the name.

3. The human form of the disease is called something else. What is it?

4. Scientists suspect that the humans infected during this outbreak acquired the disease from eating meat from diseased animals. Even when meat is well cooked, it transmits the infection. What does this say about the infectious agent?

5. These cases in Britain were not the first cases of the disease; it occurs at a low constant rate in other countries, including the United States. Although some of these sporadic cases can be traced to transplants of infected tissues, such as corneas or brain tissues, most are idiopathic. What does *idiopathic* mean?

6. Livestock control measures have been in place in Britain for several years now. Can we expect more human cases with links to the British cattle epidemic, or is it behind us? Defend your answer.

Case 2.8 Challenge

Immediately after you finished the physician's assistant program you took a job at a free clinic in the heart of the city. You always wanted to help society, especially those in the direst of life's circumstances. During your first week, you met Dwight, a 53-year-old overweight white man with several obvious problems. Dwight's feet are cracked and blistered and he has three infected toenails. (He tells you he has been homeless for various periods during the last 10 years.) There are seeping sores in the folds of his wrists and under his arms. His gums are bleeding. Dwight is here because he has been greatly confused over the past eight months and it seems to be getting worse. He has episodes of ranting and raving. He reports "feeling crazy" and being very scared. His friends brought him to the clinic and are waiting outside.

The supervising physician quickly joins you in the examining room. Together you examine Dwight thoroughly and make a plan for addressing his pressing needs for wound care. The physician starts asking him about his health history. Dwight reveals very little, saying his memory is very bad. But he talks a lot about his past sexual exploits.

1. The doctor speaks with you in the hallway. He tells you that mental illness is very common among the homeless population. Dwight needs a thorough psychiatric evaluation. The doctor is fairly sure that some of Dwight's neural symptoms are caused by a sexually transmitted infection. Which one? Caused by which microorganism?

2. If the blood test comes back positive, does it mean that Dwight can transmit the disease to others? Explain.

3. Should Dwight be treated with antibiotics to remedy his neural symptoms? Why or why not?

4. The patient's blood test came back positive. For what other infectious disease should he now be tested?

5. The doctor tells you to expect to see more of these cases in the future. But a coworker, who graduated from nursing school 10 years ago, tells you that this disease (especially its later forms) is relatively rare and is decreasing in incidence. Who is right?

The Usual Suspects

Common microorganisms causing infections in the nervous system[1,2,3]

Bacteria

Gram-positive

Clostridium botulinum

Clostridium tetani

Listeria monocytogenes

Streptococcus pneumoniae

Gram-negative

Haemophilus influenzae

Mycobacterium leprae

Neisseria meningitidis

Treponema pallidum

Fungi

Cryptococcus neoformans

Viruses

Measles virus

Poliovirus

Rabies virus

Varicella-zoster virus

West Nile virus

Arboviruses (western equine encephalitis, eastern equine encephalitis, St. Louis encephalitis, etc.)

Protozoa

Naegleria fowleria

Toxoplasma gondii

Trypanosoma species

[1] Not all of the infections appear in this chapter.
[2] Not an exhaustive list.
[3] Prions also cause disease in the nervous system.

Diseases of the Cardiovascular System

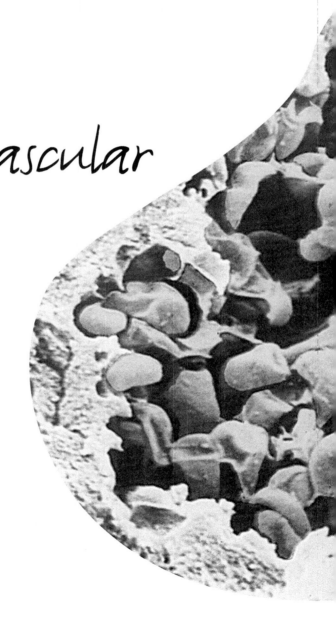

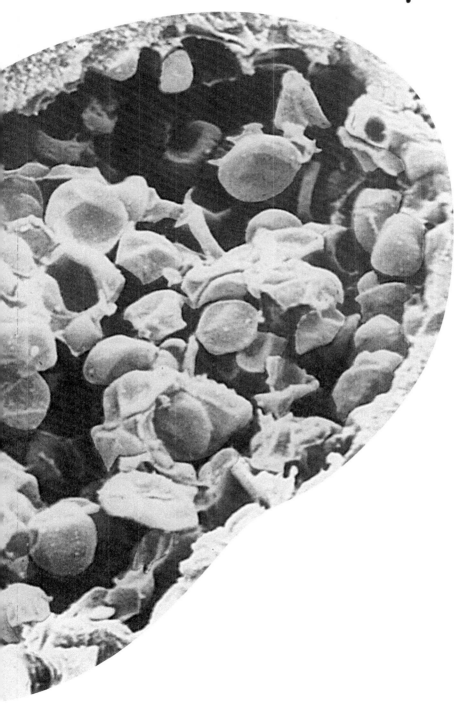

Chapter 3

Chapter Opener Three is a highly magnified photo of *Pneumocystis carinii*.
Courtesy A.B Dowsett/Photo Researchers, Inc.

Case 3.1

You are at your son's baseball game when another boy's dad experiences dizziness and nearly faints in the stands next to you. You tell him that you are a paramedic and will walk him to your car where you have your medical equipment. He reports that he has had a headache off and on since he had a tooth extracted four days ago. This evening he is feeling very weak.

His blood pressure is normal. When you listen to his heart you note that he has a pronounced **murmur.** He reports having had rheumatic fever 15 years ago. You examine his fingernails and find one that has tiny **petechial** hemorrhages under it.

1. Which cardiovascular infectious condition is this?

2. What is the most likely causative organism and the route of transmission?

3. What's the connection, if any, with rheumatic fever?

4. Why did you look at his fingernails?

5. What type of culture would a physician most likely order, and why?

6. What is the treatment? Is there a way to prevent the condition?

Case 3.2

On Christmas Eve, 2000, you were working as a clerk in a Dallas emergency room. At 3 A.M., a man and two women arrived with a screaming 6-year-old girl. The man tried to explain what was wrong, but he spoke only Spanish and you had a difficult time understanding him. The girl's mother was sobbing and you couldn't hear what she was saying. The other woman spoke a bit of English and explained to you and the nurse on duty that this was her sister's family, who had just arrived from El Salvador. The aunt did not know what was wrong with her niece but told you that the father was repeating the words for "break bone."

The nurse examined the girl and found that she had a rash and a fever of 104°F. Although the girl seemed to be in severe pain, the nurse found no bone fractures. The father, shaking his head violently, said something urgently to his sister-in-law. She interpreted his frantic statement for you, "He said it's in her blood."

1. What is your diagnosis?

2. What connection does this disease have to broken bones?

3. This is a vector-borne disease. What is the name of the most common vector?

4. What other infection is transmitted by the same vector?

5. The next night when you arrived at work the little girl seemed to be doing better. The rash had subsided and her fever had lowered. But on the third night you arrived to find that she had been transferred to intensive care after hemorrhaging internally. Is this still consistent with your original diagnosis? Explain.

6. This all sounds very bad, but you're somewhat comforted by the fact that this disease is not found in the United States. Right?

Case 3.3

You work in a small family practice in rural Virginia. A man in his early 50s comes in with a complaint of intermittent fever (102–103°F) and headache for the past two weeks. The physician examines him and takes a history. The only clinical finding is a wound about the size of a quarter on his right thumb. **Axillary** lymph nodes are swollen and tender. The man says he cut himself while skinning a rabbit three days ago. On the basis of these observations the physician prescribes streptomycin and asks the man to call if his symptoms don't improve in three days.

The physician asks you to draw blood and tells the patient he should return in four weeks for another blood sample. She says there is no need to culture the wound.

1. On the basis of the limited information above, the physician has obviously made a diagnosis. What is it? What does it look like when Gram stained?

2. What is the most likely reservoir for the causative organism in this case?

3. Why draw blood twice?

4. Why not culture the wound to look for the bacterium?

5. What are some other common infections that humans acquire from animals? (These are also known as **zoonoses.**)

Case 3.4

A 63-year-old international telecommunications executive visits your office with complaints of a high fever. The fever is not constant, but **intermittent.** When you press him for details he estimates that every three days or so he suffers these **debilitating** "sweats." He usually has headaches and muscle aches during the episodes. They keep him home from work. After half a day or so he feels better. He reports that he has experienced these episodes for about two months.

1. What is the name of the condition you suspect?

2. What should be your first question about the patient's history?

3. What is the most likely causative organism (genus and species)? Support your answer.

4. Is this pathogen eukaryotic or prokaryotic?

5. Which is the most dangerous of the species that can cause this disease? Give some details.

6. What are the two main places in the human body that are exploited by the causative organism in this disease?

7. Can this individual transmit this infection to others? Why or why not?

You've decided to work in the Peace Corps for the first two years after graduating from nursing school. Your assignment is in a rural area in South Africa. You and a coworker are setting up a clinic and encouraging women from the surrounding villages to bring their children when they are ill and to visit the clinic themselves, especially when they are pregnant.

1. In your first week you saw several children whose major symptoms were high fever, lots of sweating, and **prostration.** They all turned out to have the same infectious condition, one that you continued to see throughout your stay in South Africa. Up to half of the sick children did not survive this illness. What is it?

2. In this setting, what is the best prevention for this disease?

3. In your third month you saw a 2-year-old boy with an angry-looking rash. He was very ill with a high fever, and eventually died. His death surprised you because you thought this disease had been conquered long ago. (In the United States it is seen only occasionally, because children are vaccinated for it.) Over the course of your two-year stay you saw these symptoms in children perhaps a dozen times. Several of the children died. What is the disease?

4. Name at least two of the most common infectious conditions you should look for in adult clients in this setting.

Case 3.6 In the News

A newspaper report from Boston in the late 1990s described a growing fear among local residents. They were afraid to venture outdoors because of the increasing visibility of a particular infectious disease. The article reported that the number of people hiking in Massachusetts had recently decreased dramatically, and that many homeowners were erecting fences and spraying their yards with pesticides. Many people who dared to venture outdoors wore white clothing and tucked their pants inside their socks. In New York, there were reports of residents simply paving over their lawns, and some people gave up gardening altogether.

Although most prominent in the Northeast, similar behaviors were seen all over the country. In Montana, 10% of people surveyed felt they were at high risk for the disease, even though the Centers for Disease Control and Prevention (CDC) said that the risk was very low in that state.

1. What infectious disease do you suppose these cautious citizens were trying to avoid?

2. What determines which geographical region of the country carries risk of this disease for its inhabitants?

3. Another major disease in the United States is transmitted in a similar way. What is it, and what microorganism causes it?

4. Which regions of the country have a high incidence of this second tick-borne disease?

5. Which of these two diseases frequently has no skin manifestations at all?

Case 3.7 Challenge

Fred is a longtime patient in the family practice where you work. Typically he comes in once a year for a physical because his job involves high-steel construction work and his company requires annual checkups. However, during the past six months he has visited the office three times.

Fred first came to the office in January complaining of extreme fatigue. He had lost 15 pounds since his checkup the previous May. As part of his examination the physician ordered a standard human immunodeficiency virus (HIV) test (of the ELISA type), which came back negative. He was to come back for a repeat test in May. But he returned one month later because he was experiencing an episode of genital herpes in which the lesions had not healed in over three weeks.

The episode eventually subsided and Fred returned in May for his repeat HIV test. During this visit, he told the doctor that he was recovering from a severe respiratory infection that had bothered him for weeks. The physician drew blood; his **CD4 count** was 200 cells/ml. This HIV test (again, an ELISA) was also negative.

Two months later Fred was admitted to the hospital and a lung biopsy demonstrated *Pneumocystis carinii* pneumonia, but another HIV test came back negative. He was released after three weeks, but re-admitted with the same infection two months later. Again he tested negative for HIV. He died three days after admission to the hospital.

1. What is an ELISA test, and what does the one for HIV actually detect?

2. This patient did indeed have HIV infection, but continued to test negative. What are some possible explanations for the consistently negative test results?

3. Are any alternative tests available to clinicians for patients strongly suspected to be HIV-positive who test negative with the usual test?

4. Would you expect patients with lack of serum reactivity to have a fast or slow progression from HIV infection to acquired immune deficiency syndrome (AIDS)? Defend your answer.

5. Which of the reported symptoms are consistent with a diagnosis of HIV?

Case 3.8 Challenge

You're at the beach on Lake Michigan with your friends over spring break. The house you're staying in is a few blocks away from the beach (okay—so you're on a budget!), and the flower border around the house is overgrown with weeds. There is a tiny concrete patio next to the house where the four of you crowd to lie out in the sun when you're not at the beach.

Everything is fine until Janet complains of an insect bite on her ankle. It looks like a big mosquito bite. You rummage around under the sink in the bathroom and find a very old bottle of aloe lotion. She rubs it on the bite and you both return to the patio.

The next day Janet's ankle is very red in the area around the bite. It is hot and tender to the touch. Being nursing students, you decide not to take a chance and you drive her to the local hospital's emergency department to have it looked at. You wait there for four hours while other, more seriously ill patients are seen before you. It's your last day at the beach, and even Janet is beginning to feel it's not worth wasting the day in the waiting room. So you leave the hospital without seeing a doctor.

You go back to the house and Janet puts more aloe lotion on the bite. Then off you go to the lake. That night Janet's roommate wakes you at 2 A.M. saying that Janet is crying and sweating. When you get to her room you see that Janet looks very ill. She is covered in sweat but is shivering. She is very pale, almost blue in places, and there are red patches on her legs. You dial 911.

1. What do you suppose is happening with Janet? Is it dangerous?

2. Explain Janet's symptoms described in the last paragraph of the case.

3. What organism causes this condition?

4. When you relate the history of Janet's condition to one of the paramedics, you notice that she writes "secondary to cellulitis" on her pad of paper. What is cellulitis, and what does it mean that Janet's condition is "secondary" to it?

5. How should Janet's condition be treated at this point?

The Usual Suspects

Common microorganisms causing infections in the cardiovascular system[1, 2, 3]

Bacteria[3]

Gram-positive

Bacillus anthracis
Clostridium perfringens
Streptococcus pyogenes
Other streptococci

Gram-negative

Bartonella henselae
Borrelia burgdorferi
Brucella species
Ehrlichia species
Francisella tularensis
Rickettsia species
Yersinia pestis

Fungi

Various

Viruses

Coxsackievirus
Dengue fever virus
Ebola virus
Epstein-Barr virus
Yellow fever virus
Human immunodeficiency virus

Protozoa

Babesia microti
Plasmodium species
Schistosoma species
Trypanosoma cruzi
Wucheria bancrofti

[1]Not all of the infections appear in this chapter.
[2]Not an exhaustive list.
[3]Many bacteria can cause bloodstream infections if given access; this table lists those adapted to cause disease in this system.

Diseases of the Respiratory System

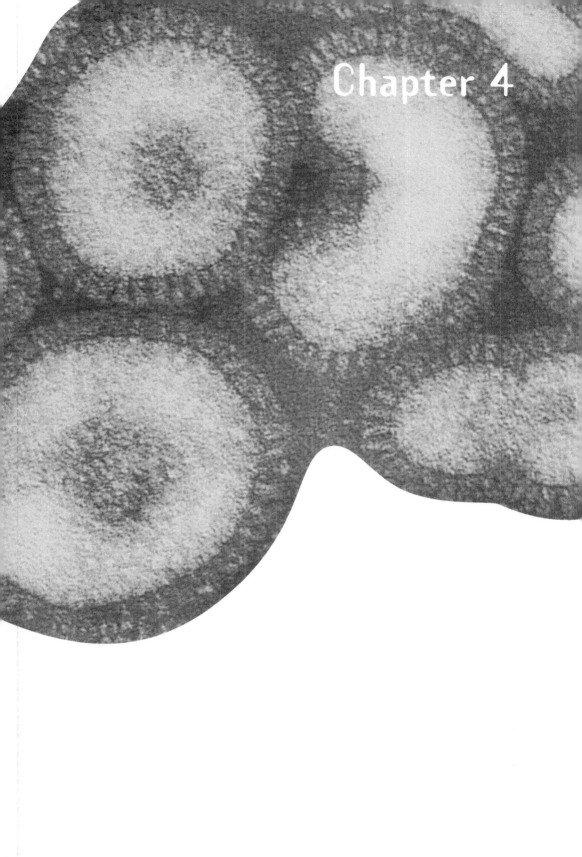

Chapter 4

Chapter Opener Four is a highly magnified photo of *Influenzavirus*.
Courtesy G. Murti/Visuals Unlimited.

Case 4.1

You are a physician's assistant at a local pediatrician's office. Five-year-old Michael is brought to the office by his father. Michael is crying and complaining that his mouth hurts. His father has been at work and does not know whether the boy has had a fever during the day. Currently his temperature is 103°F. The physician notices that Michael's breath smells rotten. Lymph nodes in his neck are swollen, and visual examination of the throat reveals a white packet adhering to the left tonsil. Much of the soft palate is red.

1. What laboratory tests are called for?

2. What types of infections are in the **differential diagnosis**?

3. Your practice has recently been overrun by sore throats and now, late in the evening, there are no supplies for performing the proper test. Should the physician prescribe antibiotics or not?

4. In deciding whether to prescribe antibiotics, should the physician be extra careful not to prescribe an unnecessary antibiotic, or be extra careful not to let a bacterial infection go untreated?

5. What are the possible **sequelae** of untreated sore throats?

Case 4.2

You and your friends are driving to the mall; it is late October. A public service announcement comes on the radio urging people to get their flu vaccinations. You are a second-year nursing student and you mention that the nursing staff at your university is holding a vaccine clinic next week.

Your friend Susan says, "I'm not getting a flu shot! Last time I did, it gave me the flu." Others in the car agree with her. But Heather asks you if it's true that the vaccine can give you the flu.

1. What should your answer to Heather's question be?

2. Heather says that because she had a flu shot last year she's going to skip it this year. Respond, with an explanation.

3. What is the difference between "**antigenic** drift" and "**antigenic** shift"?

4. What is different about the vaccine from year to year? Who decides what form it will take every year?

5. Susan wants to know why you don't have to get other vaccines annually.

6. Another friend, Dru, says that even though she had the flu shot last year she got terribly sick with the stomach flu over Thanksgiving break and missed most of her vacation. What is your explanation for this?

Case 4.3

Julie's husband Doug has not been feeling well for the past 10 days. He has congestion in his lungs and has been very tired. She talked him into going to the doctor a week ago when his temperature was 101°F. The doctor gave him some oral amoxicillin, which he took faithfully until it was gone. But she still thinks he looks sick.

Julie, Doug, and their 3-year-old daughter have just moved to Ohio from Arizona. Doug is a park ranger and he loves his job, but for the past three days he has felt too sick to go to work. His respiratory symptoms have not improved. Julie makes an appointment for him with her doctor.

1. As the physician's assistant in the office, you are the first to examine Doug. What's your tentative diagnosis, based on the history?

2. Which components of the history support your tentative diagnosis?

3. Doug's condition has not responded to the antibiotic. List two possible reasons for this finding.

4. What are some other conditions caused by this microorganism?

5. Should Julie worry that Doug can transmit the infection to her or to their daughter?

6. What precautions can be taken by other workers who may be regularly or heavily exposed to bat or bird droppings?

Case 4.4

It is mid-July. You are working as a **triage** nurse in the emergency department of a small suburban hospital in Arizona. A young, athletic-looking man in his early 20s is helped into your office by his girlfriend. He greets you and sits down, but is feverish and his breathing is labored. The girlfriend answers your questions for him. She says the symptoms began about 24 hours ago and seemed to worsen quickly. It looks like the flu to you, but the season is wrong. So you ask about the man's activities over the past week to 10 days. Nothing in this history points to an obvious **etiology** for the disease. And the girlfriend, rather defensively, adds that she is a "neat-freak" and is constantly cleaning and disinfecting the house they share. But of course, respiratory infections are very common and can be acquired anywhere. After listening to his chest you decide that it may be bronchitis or influenza. You decide to isolate him from the rest of the people in the waiting room until an examining room becomes free.

Forty-five minutes later the girlfriend comes barreling into your office. "I think he's choking!" she screams. You and the attending physician arrive at his bed where he indeed seems to be suffocating. His face is red and he is gripping his throat. The doctor calls out, "**acute** respiratory distress," and a team moves in to try to restore his breathing.

Later that evening, on your way out, you learn that the patient has died. Several days later the charge nurse tells you what the patient's lab work revealed. It identified an infection that he probably acquired a few weeks earlier while he and his girlfriend stayed in an isolated cabin his family owned but seldom used.

1. What is the diagnosis?

2. What connection does the diagnosis have with the cabin?

3. You overhear the charge nurse say to herself: "I knew there was a good reason not to clean my house." To what could she have been referring?

4. This case is from Arizona. These infections were first seen in the United States in May of 1993 in the Four Corners area of the Southwest, which includes Arizona, Colorado, New Mexico, and Utah. Can we assume that this disease is only found in the Southwest? What factors determine the places this virus might be present?

Case 4.5

Your son's best friend, Josh, has infectious mononucleosis; he hasn't been in school for two weeks. Your son and three of his friends come over after basketball practice looking for snacks, but they also want to talk to you about Josh's infection because they know you are a physician's assistant. They are all afraid to visit Josh, but they want to know when they can expect him back at practice. One of the boys asks you what causes "mono." Another one of the boys says he heard it was a form of herpes. All of the boys cringe at that one. Can you help these guys out with some information?

1. What causes mono, or infectious mononucleosis? What do you know about this agent?

2. What are the symptoms?

3. How long will Josh be out of school? Is it okay to visit him?

4. You tease the boys by saying, "Besides, by the time you're adults, all of you will have it anyway." Before they recover from that shock you add, "and some of you have it right now!" Are you just playing around with them, or are these statements true? Explain your answers.

5. Sam, the point guard on the team, says his aunt has **chronic** fatigue syndrome. "Isn't that caused by the same virus?" he asks. Is it?

Case 4.6

When you left for school this morning your 3-month-old son was wheezing a bit and he had a slight fever of 99.8°F. Your mother is watching him while you come to school to take your anatomy and physiology exam. Your pager goes off halfway through the exam. The baby's fever is rising and he is having more trouble breathing. Your mother says she is taking him to the emergency room. You rush over to the hospital. When you get there, he is in an examining room and the doctor is signing papers to admit him to intensive care. She says she suspects some kind of pneumonia. She mentions the type of pneumonia but you don't recognize the name and you are too worried about your son to pin her down at this moment. You do note that she mentions that the hospital has seen a dozen pediatric cases of this same type of pneumonia in the past week and a half.

The doctor swabs your son's nose but says the results won't be back for several days. In the meantime, they will give him supportive therapy, including an inhaled spray, but no antibacterial drugs. The doctor says that she feels sure that the child will recover, since the infection was caught very early. Nonetheless, after she leaves, your mother is frantic and indignant. She fires off the following questions to you.

1. What kind of pneumonia is it?

2. Why aren't they giving him antibacterial drugs?

3. How can the doctor be sure what's causing the pneumonia if she doesn't yet have test results?

4. What about your other child, who is 3 years old? Has she been exposed to the infection by being around the baby? Should the baby remain isolated when he comes home? Can the 3-year-old be vaccinated?

Case 4.7 In the News

One autumn in the late 1990s, a number of people became ill after working at a single building at an industrial plant in a neighborhood of Baltimore, Maryland. Their symptoms ranged from simple coughing and other respiratory symptoms to pneumonia. At least one of the 70 people reporting symptoms died.

The company voluntarily closed the building upon the recommendation of the Maryland Department of Health and Mental Hygiene. After all of the water systems at the plant were evaluated and disinfected, it reopened and no new cases were reported.

1. Health departments often have even less information than this when they have to start hypothesizing about the causative organism and its source. What is your first guess?

2. Describe the transmission characteristics of the suspected bacterium.

3. Is there a risk for a continuing community outbreak from these initial infections? Why or why not?

4. Would the health department be likely to identify this bacterium by performing routine water-screening procedures, such as serial dilution or filter collection followed by incubating on eosin methylene blue (EMB) or nutrient agar? Explain.

Case 4.8 Challenge

You have just been accepted into the nursing school at a local medical center. The program requires that you have a physical, which includes a tuberculosis (TB) test as well as the hepatitis B **recombinant** vaccine series. The nurse administering the TB skin test explains that if significant swelling occurs around the injection site, you will probably have to have a chest X-ray to determine if you are infected with *Mycobacterium tuberculosis*. One and a half days later you wake up and look at your arm, which appears swollen in an area about the size of a quarter around the skin test. It is red and tender to the touch. You're alarmed; could you have TB?

1. Why does the reaction take 36–48 hours to show up?

2. If you have a tuberculosis infection, why doesn't the whole body, or at least the respiratory tract, react when the **antigen** is injected during this diagnostic test?

3. You are referred for a chest X-ray, but the results are inconclusive. The clinic doctor prescribes a six-month course of isoniazid (abbreviated INH). You take the medicine according to the pharmacist's instructions. Six months later you are taking a medical microbiology course as part of your nursing curriculum. On the day you study tuberculosis, you suddenly realize why you had a positive skin test. It has nothing to do with a true infection, but with the fact that you were born in the Netherlands. Your family moved to the United States when you were 4 years old. What do you suppose is going on here? Discuss as fully as you can.

4. You have a friend in your hometown who is HIV-positive. When you told her about your TB scare, she said that her specialist can't use the TB skin test, even though HIV-positive people are at higher risk than the healthy population for TB. Why is the skin test not recommended for HIV-positive people?

The Usual Suspects

Common microorganisms causing infections in the respiratory tract[1,2]

Bacteria

Gram-positive

Corynebacterium diphtheriae

Stapyhlococcus aureus

Streptococcus pneumoniae

Streptococcus pyogenes

Gram-negative

Bordetella pertussis

Chlamydia pneumoniae

Coxiella burnetii

Haemophilus influenzae

Legionella pneumophila

Mycobacterium tuberculosis

Mycoplasma pneumoniae

Yersinia pestis

Viruses

Coronavirus

Epstein-Barr virus

Hantavirus

Influenza virus

Parainfluenza virus

Respiratory syncytial virus

Rhinovirus

Fungi

Aspergillus

Blastomyces dermatitidis

Coccidioides immitis

Cryptococcus neoformans

Histoplasma capsulatum

Protozoa

Uncommon

[1]Not all of the infections appear in this chapter.
[2]Not an exhaustive list.

Diseases of the Digestive System

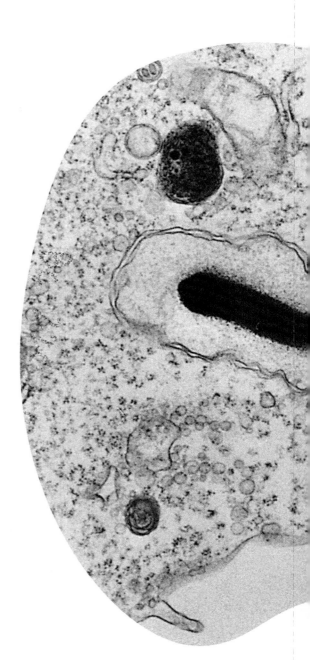

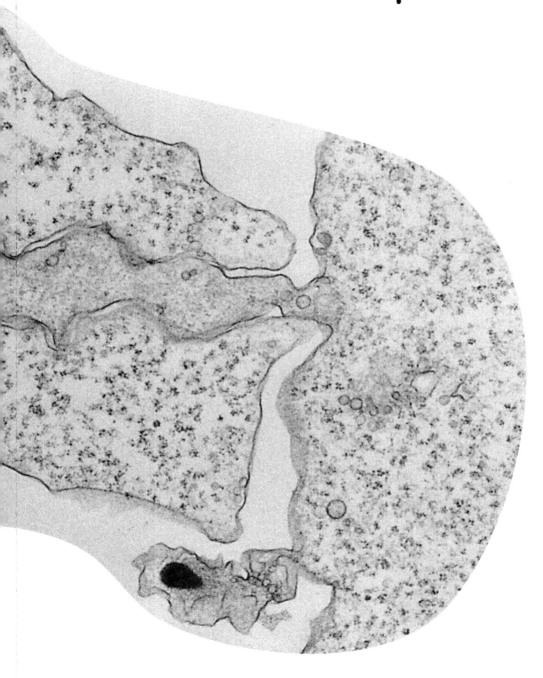

Chapter 5

Chapter Opener Five is a highly magnified photo of *Listeria monocytogenes*.
Courtesy of L.T. Tilney, P.S. Connelly & D.A. Portnoy.

You are at dinner with four of your friends. A local outbreak of *Escherichia coli* O157:H7 has been in the news. The news stories suggest that the source of the infection was unpasteurized apple cider, but the group wants to know if hamburgers are safe. They remember that there was a big outbreak of *E. coli* associated with burgers from a fast-food restaurant in the Northwest. They turn to you, since you are a nurse. You tell them to order steaks. They ask if you're buying!

1. Why steaks instead of hamburgers?

2. One of your friends acts disgusted and says she'll order a salad instead. Will this guarantee her safety? Why or why not?

3. One of your friends says that her sister gives her baby apple juice every day. Should she stop? Explain your answer.

4. What are the symptoms of *E. coli* O157:H7 infection?

5. Another friend says that his family has always eaten rare hamburgers and no one has ever gotten sick. He thinks it's all a bunch of overblown media coverage and says he will continue to eat his favorite delicacy, raw hamburger meat on crackers. What should you tell him?

Case 5.2

Last week you were on a clinical rotation at the local hospital as part of your second-year nursing program. On this rotation, your instructor took a hands-off approach and left you on your own for hours at a time. You spent most of your time hanging around at the nursing station, following nurses as they went about their duties from bed to bed, and listening to conversations between doctors and nurses about patients.

Then, one day one of the nurses who had just emerged from his fourth trip to the bathroom collapsed behind his desk. He had been losing weight and today looked especially pale. You ran to get the attending physician who was just across the hall. He took one look at the prostrate nurse and said something like "see dif" to the nursing instructor who had arrived on the scene. She replied that he had been on multiple antibiotics for the past few months in an attempt to treat a particularly nasty sinus infection.

After the sick nurse is transferred to a bed, your instructor asks you for a written report on the condition. You didn't want to admit that you weren't really sure what condition was involved here, so you figured you could look it up in your books or on the Internet at home.

1. Your Internet search of all kinds of different spellings of "see dif" yields nothing. What section of your microbiology text would likely contain the help you need? What clues lead you in that direction?

2. Now that you've found the right category of infections, can you identify what "see dif" is?

3. Your book has only a small paragraph on this infection. But now you know what to search for on the Internet to find more information. Your instructor wants you to report on the **epidemiology** of the infection. You find that it is referred to as an opportunist and this accounts for its epidemiological patterns. First of all, what is an opportunist?

4. Part of an epidemiological description of an infection involves knowing who is most often affected by it. Let's consider opportunistic infections as a group. People in which age groups are most likely to suffer symptoms from an opportunistic infection?

5. In this case the affected nurse is in his mid-30s. Is it his age or something else that predisposes him to the infection? Discuss.

6. What is the major **virulence** factor for this microorganism?

Note on using the Internet for research purposes: Always be sure that you use a reliable website, such as the Centers for Disease Control site, www.cdc.gov. You will probably find hundreds of sites from other sources, such as class notes posted by professors from various universities and student reports, as well as information from pharmaceutical companies that are marketing drugs to treat the infection. Your search may even return personal web pages of people who have suffered from the disease. This information may or may not be reliable.

Case 5.3

Your sister Pam called you last night, upset about her recent visit to the pediatrician (she has a 3-year-old son). Actually, she was upset about the discussion she had afterward with her husband, who was adamantly opposed to having their son vaccinated against hepatitis B virus (HBV). Pam called you because the doctor had convinced her that it was necessary, and indeed routine, to vaccinate young children. Her husband believes that hepatitis B is mostly acquired through sexual contact and drug use and that it's ridiculous to vaccinate a 3-year-old. Pam wants your advice before continuing this discussion with her husband.

1. First of all, is Pam's husband correct about transmission of the virus? Elaborate.

2. How severe is this infection for young children?

3. Pam says she'll also remind him that in the last year the newspapers have reported at least three hepatitis outbreaks traced back to restaurants. Respond to her statement.

4. While you're on the phone with Pam, her husband comes home from work. He hears your conversation, and says in a loud voice, "That vaccine is not safe! It's one of those genetically engineered things!" What can you tell Pam about how the vaccine is made, and whether it is safe or not?

Case 5.4 In the News

One summer in the late 1990s, a group of tourists from the United Kingdom became ill after they all stayed at the same hotel in Greece. Epidemiologists conducted surveys among all the people who had stayed at that hotel during the two-and-a-half-week period in which people were reporting their illnesses. They did this in an attempt to determine the cause of the symptoms, which were primarily diarrhea and nausea. They surveyed 239 people; 224 of them reported having been ill while they were still on vacation or shortly after their return. Their diarrheal symptoms lasted 10–15 days.

Seventy of the 224 people who reported illness were classified as having definite cases of gastrointestinal disease. A case was called "definite" when a pathogen could actually be recovered from their stool. Of these, the vast majority tested positive for one particular microorganism.

1. Microscopic analysis of the stool samples revealed the presence of small oval-shaped structures, with defined outer walls and two to four nuclei inside that looked like seeds. What is your diagnosis?

2. What organisms should be included in the **differential diagnosis** of this infection?

3. What feature of the symptoms suggests that the causative organism is not likely to be *Staphylococcus aureus*?

4. Epidemiologists interviewed the patients about their vacation activities and food intake to try to identify the environmental source of the infection. There was no relationship between illness and a person's attending one of the scheduled children's activities at the hotel. Only two types of food available in the dining room seemed to be associated with the illness: raw vegetables and salads. There was also a statistically significant relationship between illness and having consumed orange juice made from a mix (with hotel water). So what was the likely source?

5. Why would an epidemiologist even ask about a person's attendance at children's activities?

6. Are there any symptoms that would help to distinguish this kind of diarrheal illness from others?

Case 5.5 In the News

On Christmas Eve a few years ago, the Ohio State Health Department announced that two elderly people had died during the previous six weeks, apparently after ingesting tainted meat. Ten additional nonfatal cases were reported in the state during this period.

The state epidemiologist was aware of a national outbreak of a disease with the symptoms seen in these cases. The symptoms included fever and muscle aches and often diarrhea and nausea. Occasionally, the central nervous system was affected, resulting in confusion, stiff neck, headache, and convulsion.

The nationwide outbreak affected approximately 40 people, with a particularly high infection rate in pregnant women and a significant number of deaths among fetuses. The Centers for Disease Control and Prevention (CDC) issued a list of people who were at particular risk for the disease. These included pregnant women, newborns, people with weakened immune systems, and the elderly.

The CDC and the U.S. Department of Agriculture suspected prepackaged meats, such as hot dogs and cold cuts, as the source of the outbreak. A recall of meat processed at a particular plant in Michigan was instituted.

1. What is the most likely causative microorganism in this outbreak?

2. Why is this infection associated with processed meats, but usually not with hamburger or cuts of meat including pork, beef, or chicken?

3. Epidemiologists describe a microorganism's **pathogenicity** as the proportion of people who become ill after being exposed to the microorganism. (An infection that is **subclinical** in most people who acquire it is considered to have low pathogenicity.) After considering the types of people at high risk for the disease, would you suppose that this organism has high or low pathogenicity? Explain your answer.

Case 5.6

You are working as a receptionist at the only family practice in a small town in Idaho while you are studying to become a physician's assistant. On a Saturday morning you are the only office worker there when a call comes in from a local church. The congregation is hosting a family that moved to the United States from Peru six weeks ago and is helping them find housing and work. In the meantime, the family is staying at a church-owned house and relying heavily on church members for help negotiating this new country and for translation while their English is still sketchy.

The woman on the phone identifies herself as Leslie, a church member. She seems distraught. She says that the mother of the young family became ill yesterday and seems extremely ill now. Her symptoms started out as stomach cramps and quickly progressed to a very watery diarrhea. You hear moaning in the background and Leslie tells you that the patient is pointing to her calves and crying. You ask Leslie how many stools the sick woman has had in the last 12 hours. She replies that it is almost constant and that the woman can no longer leave her bed at all.

When asked, Leslie says there is no blood in the excreta. It is very clear with lots of little white flecks in it. You put her on hold and run down the hall to the examining room where a physician is doing a well-baby check.

1. When the doctor opens the door you whisper that you think there's a case of _____ on the phone.

2. The doctor's eyes widen and she asks you how you came to that conclusion. What is your reply?

3. Why was the doctor initially dubious about your diagnosis and why does the patient's recent immigrant status convince her that your diagnosis was correct?

4. The doctor asks you to tell Leslie to call 911. The sick woman should be transported to an emergency room right away and the doctor will call ahead and meet her there. What is the first intervention likely to be performed when the patient arrives?

5. The incubation period for this disease is one to four days. Can you think of any way that the young mother could have been infected so recently even though she has been in this country for six weeks?

6. The next day you ask the doctor about the patient's status. She says that currently the patient is receiving a course of the antibiotic ciprofloxacin, though it won't help her. Why won't it help her and why was it prescribed if it won't?

Case 5.7

You went to get a haircut yesterday and your stylist was having a conversation with another stylist in the shop. The second stylist said that her live-in boyfriend of three years just got a blood test and discovered he has hepatitis C. Then she shrugged her shoulders and said her boyfriend has never had any symptoms so she wasn't going to worry about it.

1. After the second stylist walks away, your stylist asks you about hepatitis C. Her first question is, "Is it serious?" Answer this question as thoroughly as you can.

2. How is it transmitted?

3. Can she be vaccinated against it?

4. Your stylist has heard of hepatitis A and hepatitis B, but never hepatitis C. Is it new? Explain.

Case 5.8 Challenge

When you arrived at work in the intensive care unit this morning, you learned that a patient with Guillain-Barré syndrome had been admitted. He is a 45-year-old poultry farmer.

He is on a respirator and has **bilateral** paralysis of his legs. You remember that Guillain-Barré syndrome is a result of the immune system attacking the peripheral nervous system.

1. What leads to Guillain-Barré? What would you look for in the patient's history?

2. Considering this patient's profession, what type of condition do you suspect as the **precipitating** event?

3. Is Guillain-Barré often fatal?

4. What about the original infection, which you identified in question 2? Is it common or uncommon?

The Usual Suspects

Common microorganisms causing infections in the digestive tract[1,2]

Bacteria

Gram-positive

Bacillus cereus
Clostridium difficile
Clostridium perfringens
Listeria monocytogenes
Staphylococcus aureus
Streptococcus mutans

Gram-negative

Campylobacter jejuni
Escherichia coli O157:H7
Other *Escherichia coli*
Helicobacter pylori
Salmonella species
Shigella species
Vibrio cholerae
Yersinia enterocolitica

Viruses

Epstein-Barr virus
Hepatitis A
Hepatitis B
Hepatitis C
Hepatitis E
Mumps virus
Rotavirus

Fungi

Candida albicans

Protozoa

Cryptosporidium
Entamoeba histolytica
Giardia lamblia

[1]Not all of the infections appear in this chapter.
[2]Not an exhaustive list.

Diseases of the Genitourinary System

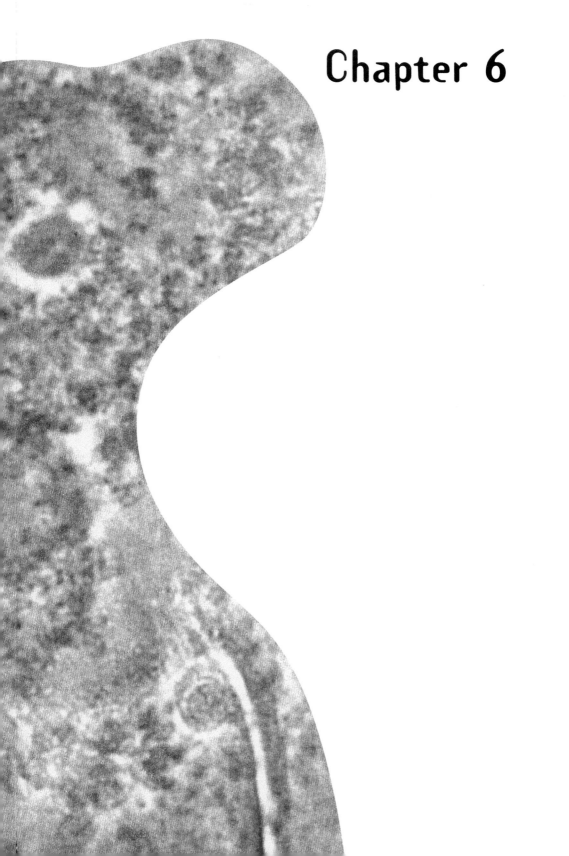

Chapter 6

Chapter Opener Six is a highly magnified photo of *Gardnerella vaginalis*.
Courtesy of the CDC.

Case 6.1

As part of a community service requirement in your second-year nursing program, you are volunteering at the local clinic for sexually transmitted diseases (STDs). At the clinic, you are responsible for conducting intake interviews. When patients arrive, they fill out a questionnaire and then you take them to an office to go over their answers with them.

Your first patient is a 22-year-old woman. On the questionnaire, she lists her chief complaint as "pain in her belly." You wonder why she has come to the STD clinic for belly pain and speculate it is because this clinic is the only one in town that is free. You excuse yourself and ask the head nurse if it's okay to continue with this client's questionnaire. "Why wouldn't it be?" is his answer. You say you don't think belly pain points to a sexually transmitted disease. The nurse chuckles and says it is one of the most frequent complaints they have in this clinic.

1. What is the association between belly pain and STDs?

2. What causes the belly pain, exactly?

3. Is the belly pain a serious sign?

4. What diagnostic tests are called for? Treatment?

5. You look further down on the questionnaire and see there is a question about the client's recent sex partners. She answered that she has had relations only with her boyfriend during the past 12 months. Does the boyfriend need information or treatment? Explain.

Case 6.2

You met your wife Becky at the local hospital where you are a nurse and she works in medical records. You were married 18 months ago and now she discovers she is pregnant. You are both extremely happy and you go together to the obstetrician for her first prenatal visit. Everything looks good; she is seven weeks into a normal pregnancy. About 10 days later you come home from work and she is waiting for you in the kitchen, looking very upset. She tells you that the obstetrician's office called this afternoon with the news that she has tested positive for gonorrhea.

She says it must mean that you have been unfaithful to her, since she knows for sure that she has had no relations with anyone else since you were married. She is so distraught that she will not listen to what you have to say. She packs a suitcase and drives to her mother's house across town.

You believe Becky without question. At the same time you know that you have also been faithful.

1. How is this infection possible if both of you have been monogamous for at least two years?

2. What about the immediate problem? Should Becky be treated, even though she is pregnant? Discuss.

3. Is penicillin the best treatment for gonorrhea? Why or why not?

4. Once these facts are explained to Becky, she calms down. But you wonder if there is a shadow of doubt about your fidelity in her mind. You wonder how other couples fare—particularly those who may have less trust in one another, or don't know how to access information about STDs. Speculate about why Becky's physician did not explain all the possibilities.

You are working the night shift on The Answer-Line, a telephone service provided by a health management organization for its enrollees. You receive a call from a 26-year-old woman. She has been experiencing painful urination for the past 24 hours. This is the first time she has had this condition and she describes herself as otherwise healthy.

1. What are some of the first questions you should ask?

2. The patient reports a fever of 101.5°F and a nearly constant urge to urinate, though she often voids little or no urine. What is your preliminary diagnosis?

3. There is a certain symptom that she has not mentioned. What is it and why is it important that you ask her about it?

4. What is the most likely causative organism for this condition?

5. What is the route of transmission of this organism?

6. What are some other causative organisms for this condition?

Case 6.4

Your roommate Jane complains to you that she has had intense vaginal itching for the past day and a half. (She often asks you medical questions because you are nearly finished with your nursing degree.) She says the same symptoms crop up from time to time and that she buys over-the-counter antifungal creams that seem to take care of them after a few days. "I guess they are yeast infections," she says. "But I don't know where I keep getting them from." She adds that she hasn't had sex with anyone since her junior prom, three years ago. When you ask her how frequent the episodes are, she says about once every six weeks or so.

1. What is the causative organism of vaginal yeast infections? Where is Jane "getting them from"?

2. What conditions could **predispose** a woman to such frequent yeast infections?

3. Should Jane continue to self-medicate for her yeast infections or should she see a doctor? Please explain.

4. Are there any possible serious consequences of vaginal yeast infections?

Case 6.5

A pregnant woman arrives at your practice because she has noticed a **copious** vaginal discharge and is worried that it may indicate problems with her pregnancy. After a pelvic examination, the physician says there is a whitish, smooth coating on the walls of the vagina. Microscopic examination of vaginal fluids reveals the presence of "clue cells." The physician, a rather gruff type, writes "bacterial vaginitis" on the chart, prescribes an antibiotic, and moves on. The patient turns to you for answers.

1. What usually causes bacterial vaginitis (often called BV)?

2. What are clue cells? What bacterium is associated with clue cells?

3. Is BV a dangerous condition? Explain.

4. Is BV a sexually transmitted disease? Elaborate.

5. What other diseases are in the **differential diagnosis** of a woman with copious vaginal discharge?

6. Is BV treatable? If so, with what? What is the likely outcome of treatment?

Case 6.6

Your best friend, Jack (a 30-year-old investment banker), has had a steady girlfriend for the past six months. He has avoided having sex with her because she told him she has genital herpes. You remember the day that she told Jack about it; he came over to your house very upset and the two of you talked for hours about what that meant for Jack. He thought about breaking up with her because he couldn't see how they could have a long-term, intimate relationship. But finally he decided that he did love her and they would figure something out.

Now she wants to take the relationship to the next level, a level that includes sexual relations. She told Jack that they would do this only if she were lesion-free and that if he wore a condom, they would be fine.

Jack is skeptical. He comes to you for advice.

1. Are they safe if she does not have lesions at the time of their intercourse? Why or why not?

2. Whether Jack's girlfriend has lesions or not, if he uses a condom he will be protected, right?

3. If Jack were the one with herpes and his girlfriend was uninfected, would his use of a condom completely protect her? Explain.

4. What would you say is the safest way for Jack and his girlfriend to have intercourse?

5. What about those new drugs Jack has heard about on TV? Can his girlfriend take those and cure herself? Or at least avoid infecting him? Give some detail.

Case 6.7 Challenge

Your mother told you a story about a 14-year-old patient that she saw in the early 1980s when she was a nurse in a gynecologist's office.

Your mother's first contact with the young girl was after she vomited in the waiting room. She told your mother that she started feeling ill the night before. She had been having unusually heavy menstrual bleeding and reported having a fever earlier that morning. The young patient complained of chills and had a diffuse rash on her arms and legs.

A physician arrived on the scene and he and your mother helped her back to an examining room. Your mother checked her temperature and her blood pressure while the doctor asked her some questions. His first question was whether her neck was stiff or painful. She answered no, but the doctor ordered a **lumbar puncture** anyway. The patient was starting to look dizzy and her blood pressure was low: 90/70. The doctor asked her if she had ever had sexual intercourse and the patient answered that she had not. When the patient's mother came in from parking the car the doctor asked if her immunizations were up to date. The mother confirmed that they were.

Your mom added that the other peculiar thing about this patient was that several days after she was admitted to the hospital, the skin on the palms of her hands began to slough off.

1. What kind of infectious diseases come to mind when a widespread rash is seen as the primary complaint? (Hint: Why had the doctor asked about her sexual history? Why did he ask about her immunizations?)

2. Her rash was diffuse, with well-separated bumps that were **maculopapular.** Was it likely to be chicken pox? Why or why not?

3. The cerebrospinal fluid obtained from the lumbar puncture was clear—no evidence of bacteria. Another infection was ruled out. Which one?

4. The doctor then asked the patient about her menstrual history and practices. She began menstruating at the age of 12 and reported that her last period began four days ago. She reported that she mainly uses tampons during her period. What infection do you think the doctor had in mind in asking about menstruation? What do you know about the infection in question?

5. Your mother says that if you see a patient with these symptoms once you start your practice as a physician's assistant, it is less likely to be the same infection. Why?

Case 6.8 Challenge

A sophomore named Michelle **presents** at the college health service (where you are observing for the day) with fever (105°F), **malaise,** headache, and pain in her genital region that is severe upon urination. The physician examines her genital region and finds four blisterlike lesions on the outer **labia,** each 2–3 mm in size. The lesions are filled with clear fluid; there is no sign of bleeding from them. While you are taking her history, Michelle reports that she has had two successive abnormal Pap smears in the past year. After the first abnormal Pap, she was treated with an antiprotozoal drug. She reports taking the full course of the antibiotic.

1. What questions about the patient's behavior should the physician ask during the history?

2. What is your **presumptive diagnosis** based on the facts presented? What other conditions might be in the **differential diagnosis**?

3. Why do you suppose the patient was treated with an antiprotozoal drug after her first abnormal Pap?

4. Of what importance is the patient's history of abnormal Pap smears?

5. What tests should be ordered to confirm the presumptive diagnosis?

The Usual Suspects

Common microorganisms causing infections in the genitourinary tract[1,2]

Bacteria

Gram-positive

Staphylococcus aureus

Staphylococcus saprophyticus

Streptococcus pyogenes

Gram-negative

Chlamydia trachomatis

Escherichia coli

Gardnerella vaginalis

Haemophilus ducreyi

Leptospira interrogans

Neisseria gonorrhoeae

Treponema pallidum

Viruses

Herpes simplex virus

Human papillomavirus

Fungi

Candida albicans

Protozoa

Trichomonas vaginalis

[1]Not all of the infections appear in this chapter.
[2]Not an exhaustive list.

Glossary

Acute of short duration, rapid and abbreviated in onset, in reference to a disease

Anatomical diagnosis the identification of the physical site of symptoms

Antigen a substance, usually protein or carbohydrate (as a toxin or enzyme), capable of stimulating an immune response

Antigenic capable of stimulating an immune response

Axillary relating to or located near the axilla, which is the armpit

Bilateral having, or relating to, two sides

CD4 count number of CD4 or T$_{helper}$ cells in a milliliter of blood; used as an indicator of progress of HIV infection

Chocolate agar a semisolid medium made by adding heated blood to nutrient agar

Chronic of long duration or frequent recurrence, referring to a disease or ailment

Communicable capable of being transmitted from one host to another; contagious

Copious large in quantity; abundant

Debilitating describes a condition that saps the strength or energy, severely reducing patient's function

Debride to cleanse by surgical excision of dead, devitalized, or contaminated tissue and removal of foreign matter from a wound

Dermatophytes types of fungi parasitic upon the skin or skin derivatives (hair or nails)

Differential diagnosis a list of diseases or conditions presenting similar symptoms

Epidemic an outbreak of a contagious disease that spreads rapidly and widely; incidence of disease above normal levels

Epidemiology the branch of medicine that deals with the study of the causes, distribution, and control of disease in populations

Etiological diagnosis the determination of the root cause of a disease; if disease is infectious this involves identifying the causative microorganism

Etiology the cause or origin of disease

Intermittent stopping and starting at intervals

Labia   any of four folds of tissue of the female external genitalia

Localized   confined to a specific area of the body; not **systemic** or generalized

Lumbar puncture   the insertion of a hollow needle beneath the arachnoid membrane of the spinal cord in the lower back region to withdraw cerebrospinal fluid or to administer medication

Maculopapular   describes a small, solid, usually inflammatory elevation of the skin that does not contain pus; area of discoloration on the skin caused by excess or lack of pigment

Malaise   a vague feeling of bodily discomfort, often accompanying illness

Murmur   an abnormal sound, usually emanating from the heart, that sometimes indicates a diseased condition

Oral history   portion of a clinical examination in which the health care provider asks pertinent questions about the patient's symptoms and past and present exposures and conditions

Pathogenicity   ability to produce disease; with microorganisms calculated as the proportion of people who become ill after being exposed to the microorganism

Petechial   marked by a small purplish spot on a body surface, such as the skin or a mucous membrane, caused by a minute hemorrhage

Precipitating   leading to, as in *precipitating event,* the event that led to the current condition

Predispose   to make susceptible or liable

Present   (in clinical practice) to make oneself present to a health care provider

Presumptive diagnosis   the best guess about the nature or cause of a disease before definitive answers are available

Prostration   total exhaustion or weakness; collapse

Recombinant   pertaining to a genome that is genetically engineered; an organism that has foreign genetic material incorporated into its genes

Sequelae (sing.: sequela)   pathological conditions resulting from a disease

Subclinical   not manifesting characteristic clinical symptoms; referring to symptoms that are unnoticeable

Systemic affecting the entire body or an entire organism

Triage a process for sorting injured or ill people into groups based on their need for immediate medical treatment

Virulence degree of damage that can be produced by an infectious agent

Zoonosis (pl.: zoonoses) a disease of animals that can be transmitted to humans

Index